Photo Intro
to
Metal Detecting

Vince Migliore

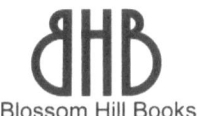

Blossom Hill Books

Title ID 3462142

Title: **Photo Intro to Metal Detecting**
An Image-based Primer

Description: ***Photo Intro to Metal Detecting*** is a photographic introduction to key concepts in metal detecting, and a guide in choosing the right detector.

ISBN is 1453638261
EAN-13 is 9781453638262
Primary Category: Sports & Recreation / General

Country of Publication: United States
Language: English
Search Keywords: metal detecting treasure hunting
Author: Vince Migliore
Blossom Hill Books
113 Sombrero Way
Folsom, California 95630 USA
Reorder: https://www.createspace.com/3462142
First Edition, July 2010

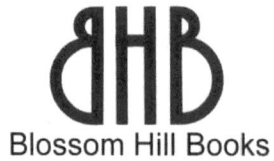

Blossom Hill Books

Table of Contents

Photographic credits:
Front cover: A. Common coins and junk with similar phase shifts. B. Scanning the ground with
overlapping swings; Nick Meinzer. C. Phase shift scale on the face plate of the White's MXT detector.
D. Typical Topo Map from the US Geological Survey.
Thanks to Joe and Phil Gouff for the Civil War relics appearing on the rear cover and in Figures 27 and 40.

Thanks to members of the Sacramento Valley Detecting Buffs for multiple pictures.

This book is dedicated to the generous people in the metal detecting community who so
willingly give their time and knowledge to the furtherance of the hobby.

Who

Who benefits from this book?
- Anyone in a hurry to learn about metal detecting
- People who are visually oriented and prefer illustrations over words
- Beginners who need a quick and simple guide to the hobby

What

What will I learn?
- Important technical terms necessary for choosing a detector, such as Discrimination, Target ID, and Phase Shift
- How detectors work and what all the dials and displays mean
- An overview of the different types of metal detectors

Where

Where can I go for more information?
Check out the resources listed at the back of the book. This includes:
- Best Bets for Beginners – Detector choices
- Hobbyist magazines and online resources
- A comprehensive list of manufacturers and suppliers

INTRODUCTION

Metal detecting is a great hobby! It gets you exercising in the great outdoors, meeting new people, and finding all kinds of treasures.

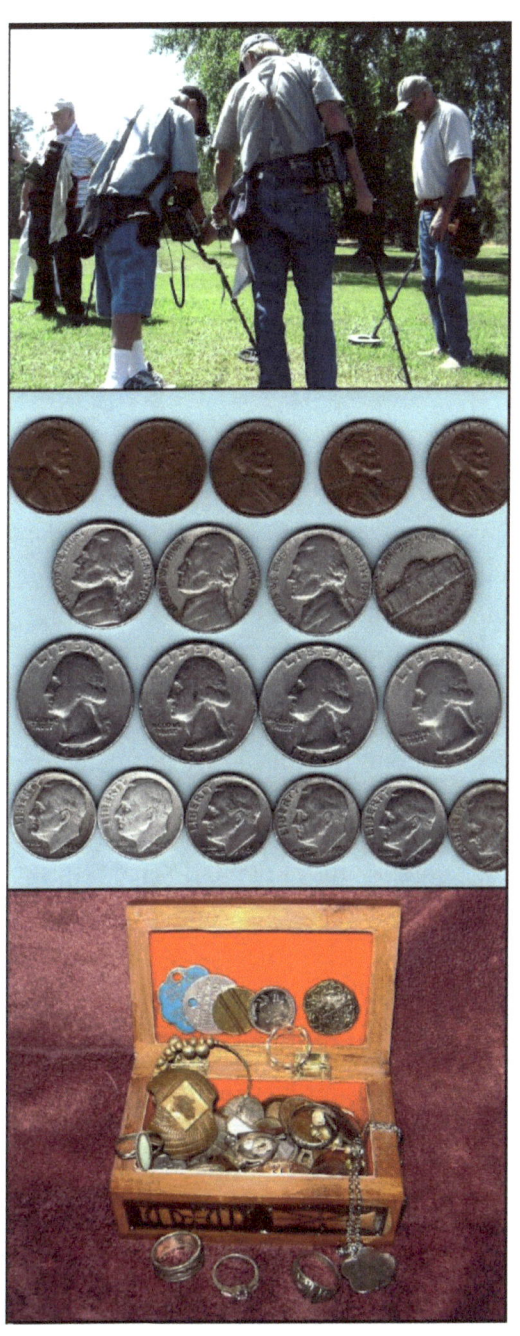

Figure 1.
Metal detecting is often a group sport, especially if you join a club. Friends and other club members can help get you started quickly.

Figure 2.
In this hobby you find all sorts of coins and tokens.

Figure 3.
Finding rings, jewelry, and other metal objects is as much fun as finding coins.

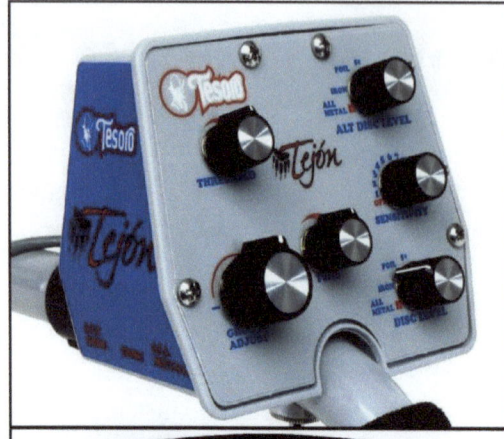

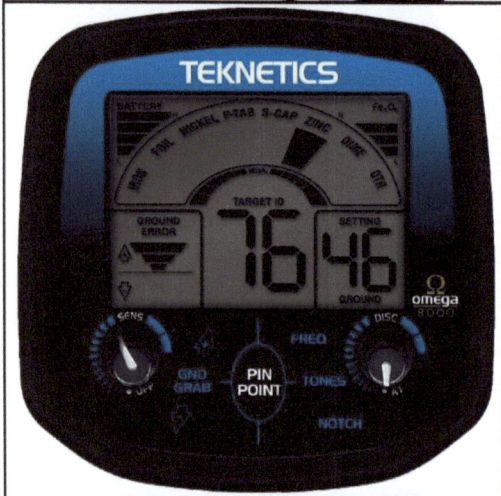

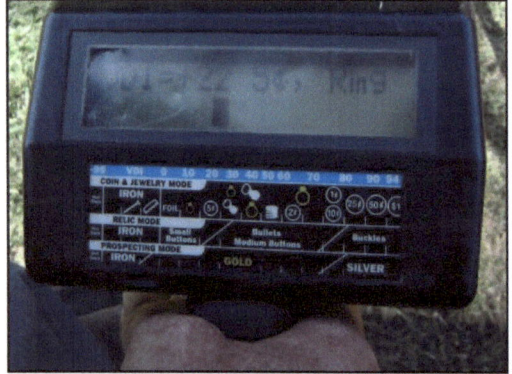

GETTING STARTED

Figure 4.
There are many different kinds of detectors. Some have dials, like the Tesoro Tejon. Some have displays. Some have both dials and a display screen. What do they all have in common?

Figure 5.
The main feature of the modern metal detector is its ability to tell the difference between a *valuable* target, such as a coin or a gold ring, and a piece of junk, like a nail or a bottle cap. That feature is called **DISCRIMINATION** – an important word in metal detecting. The black bar at the top of this Teknetics detector indicates a 'ZINC' penny. In 1982 the US switched from copper to copper-coated zinc pennies. Coated coins are called "clad."

Figure 6.
Here's what the Discrimination scale looks like on a popular model, the White's MXT. Notice that most coins are on the far right, while nails and bottle caps are on the left. Nickels, pull-tabs, rings, and gold are near the middle.

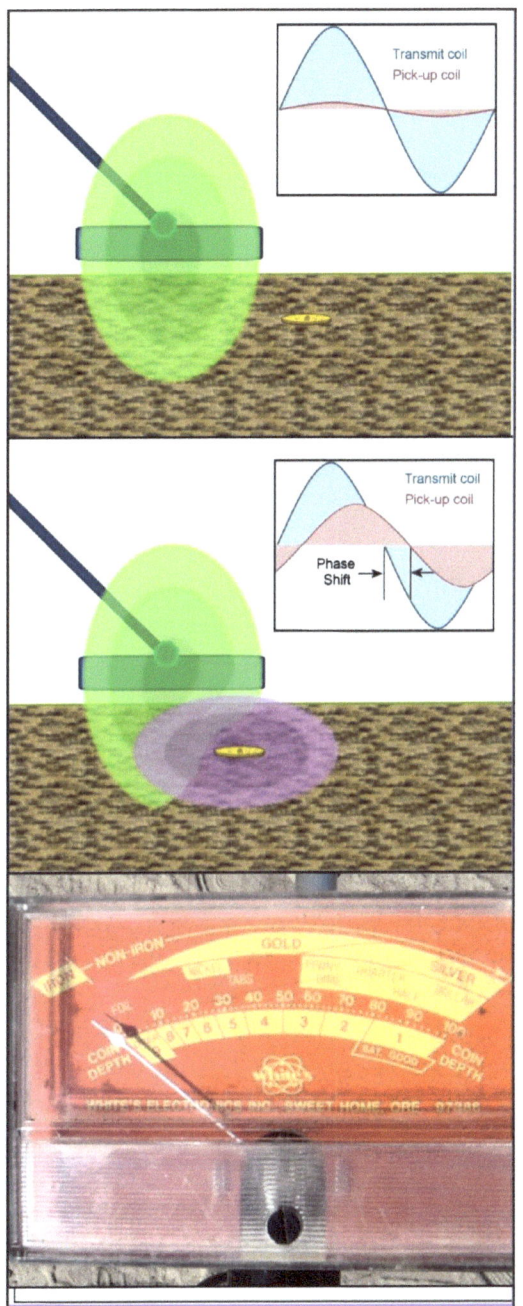

Figure 7.
Here's how Discrimination works. The coil on the metal detector actually has two loops inside. One creates a magnetic field and the other listens for a response. If there is nothing in the ground beneath the transmit coil, then there is very little response in the pickup coil.

Figure 8.
If there IS a coin or a silver ring in the ground, the magnetic field from the transmit coil creates a small electric current in the target. This current in the target, in turn, creates a current in the pickup coil that is out of synch with the transmit signal. This represents a **PHASE SHIFT**, which the detector measures. The Phase Shift helps identify the coin.

Figure 9.
A silver coin creates a large phase shift. A piece of junk, like iron, produces a small phase shift. Many detectors link the phase shift to the sound you hear, so a rusty nail produces a low grunt, a nickel creates a mid-range whistle, and a dime emits a high-pitched squeal – a sound you will learn to love! Shown here is an older detector, a Whites 6000 Di, which uses a meter to show phase shift. Most of the coins are in the upper right of the dial, indicating a large phase shift.

3

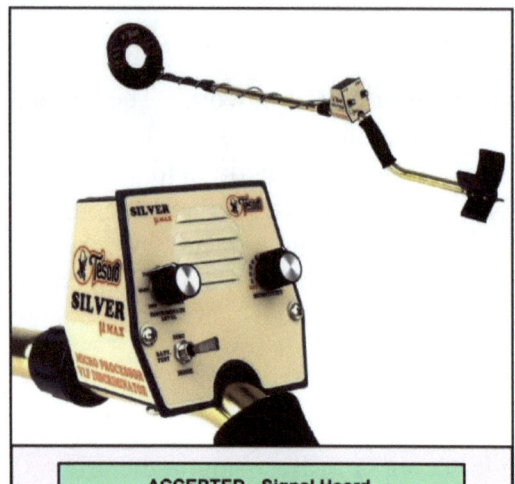

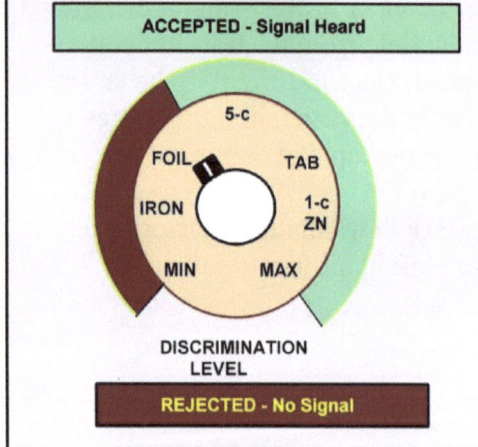

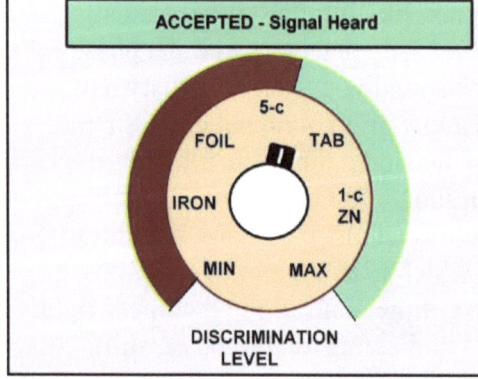

EXAMPLES OF DISCRIMINATION

Here's what Discrimination looks like on different detectors.

Figure 10.
The first example is the Tesoro Silver Umax. You can see a speaker and the Discrimination dial. This is a SOUND ONLY machine. This detector uses sounds to tell the difference between junk and coins.

Figure 11
When the Discriminator is set to just below the 5-cent mark and you hear a signal, it can be a nickel or anything higher on the phase shift scale. That includes a pull-tab, a penny, a dime, a quarter, and so forth.

Figure 12.
If you move the Discriminator dial up past the 5-cent mark, but below the Tab, and the signal goes away, that means you just "discriminated out" the nickel. The target is most likely a nickel. Here's the key: by adjusting the Discrimination dial, to see when the signal goes away, you can tell what object was just discriminated out. That's a basic method of target identification.
If you are confused, skip it for now and look at the next model.

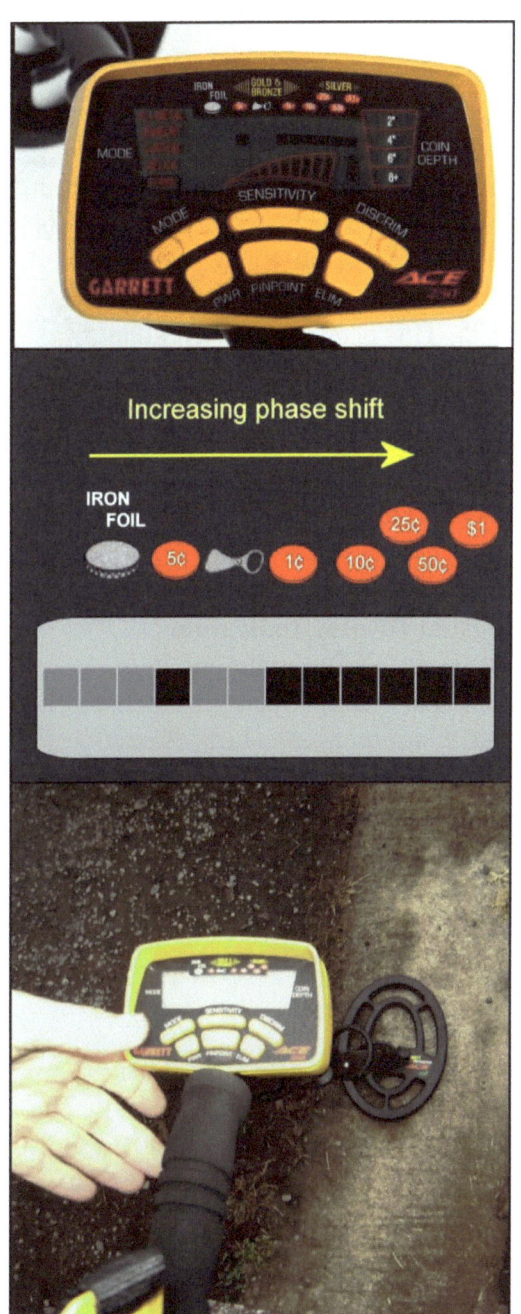

Figure 13.
Here is what Discrimination looks like in the Garrett Ace 250. This control box shows little pictures, icons, of the target. Compare this scale to the meter scale in Figure 9, and the Discriminator dial on Page 4. Now you can see what the Phase Shift scale looks like, and how it's related to another important concept: **Target Identification**, or TID.

Figure 14.
Here's the Garrett Ace 250 set to Coin Mode. Notice how the active boxes (in black) correspond to different coin values. Study this image! It's your key to understanding Phase Shift, Discrimination, and Target Identification. You can see bottle caps and iron are "notched out" as well as the pull tabs between a nickel and a penny. Those segments are grayed out.

Figure 15.
The Garrett Ace 250 is a popular model. Phase Shift and Discrimination is made possible by Very-Low Frequency (VLF) circuit technology, and you will sometimes hear people call these machines VLF detectors.

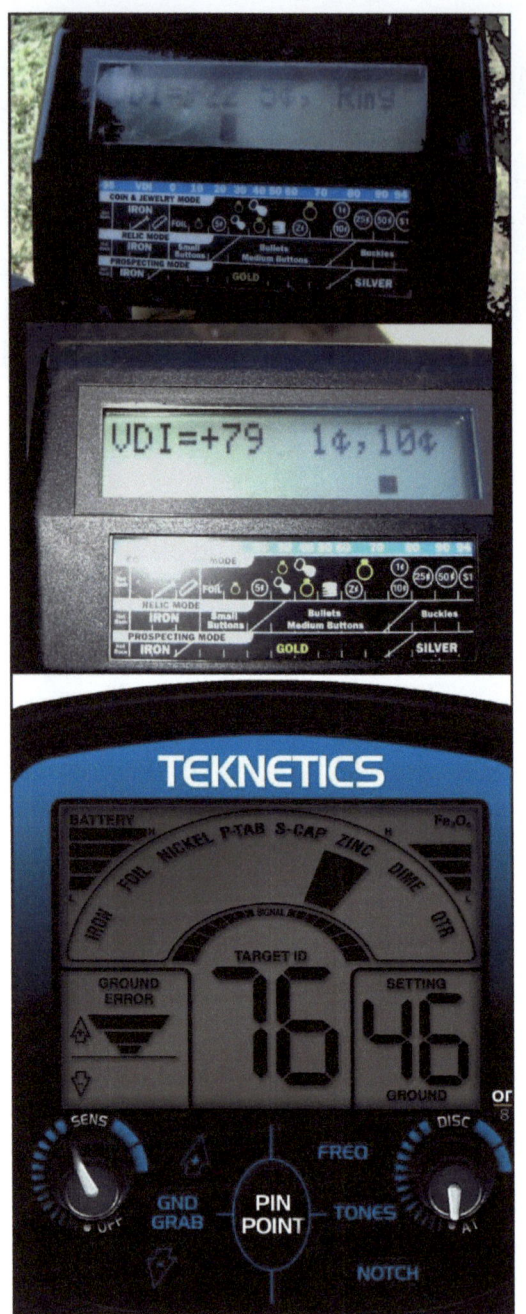

Figure 16.

Advanced metal detectors take Discrimination to a more precise level. In Figures 13 and 14, you can see that the Discrimination spectrum, the display, is broken up into ranges, or slots. In the White's MXT, instead of categories, you have individual numbers ranging from -95 to +95. This provides a much greater degree of detail than the category detectors. Here we see the number +22 corresponds to a nickel or a ring. Can you see where that falls on the blue scale in at the center?

Figure 17.

The specific values on the -95 to +95 scale are called **Visual Display Indication** (VDI) numbers. Here a VDI number of 79 corresponds to a copper penny or a clad dime, which have similar phase shifts. Sometimes there is cross-over, or mix-up between coins with similar phase shifts, but most of the time with these detectors you can tell a copper penny (VDI=77) from a zinc penny (VDI=67), or a clad dime (VDI=80) from a silver dime (VDI=81).

Figure 18.

Here's another detector that uses VDI numbers: the Teknetics Omega detector. The big "76" in the center indicates this is one of the more advanced VDI type detectors, so Target Identification is more accurate.

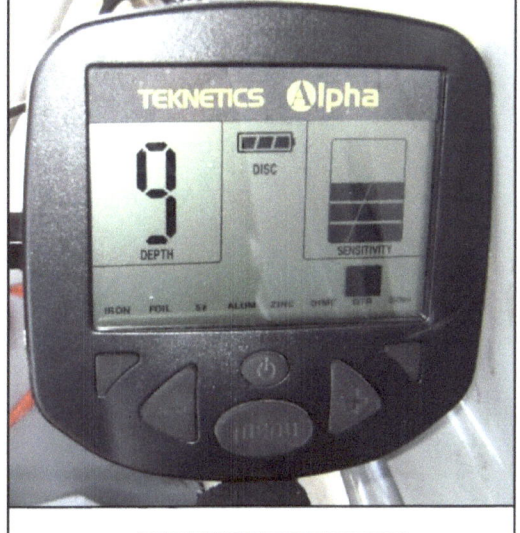

CONGRATULATIONS! If you have a grip on the ideas on the previous pages, then you've mastered the three most difficult concepts in metal detecting: Discrimination, Phase Shift, and Target Identification (TID).

Figure 19.
You can see that displays are a little easier to understand than figuring out the discrimination dial by itself. Here's the Teknetics Alpha with the discrimination categories at the bottom showing a quarter.

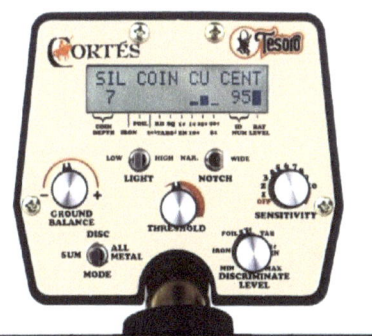

Figure 20.
Here's another detector display which uses VDI numbers. Can you see the "95" on the right side of the display? This is most likely a 25-cent coin.

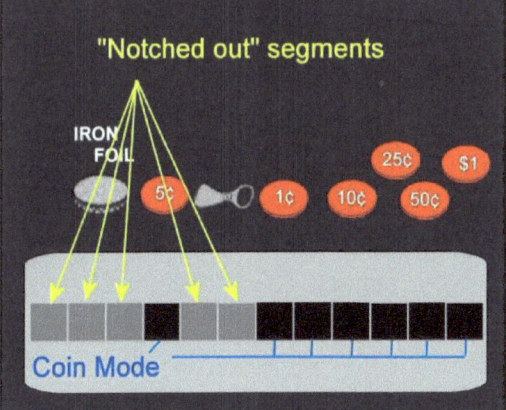

Figure 21.
Let's look again at the Ace 250 display. Notice there are 12 categories of discrimination. You can turn any one of these slots, or notches, on or off. This is called **NOTCH FILTERING**. Notice too that those pesky pull-tabs fall right in the middle of the phase shift scale. Notch Filtering lets you ignore the pull-tabs when you are searching for coins.

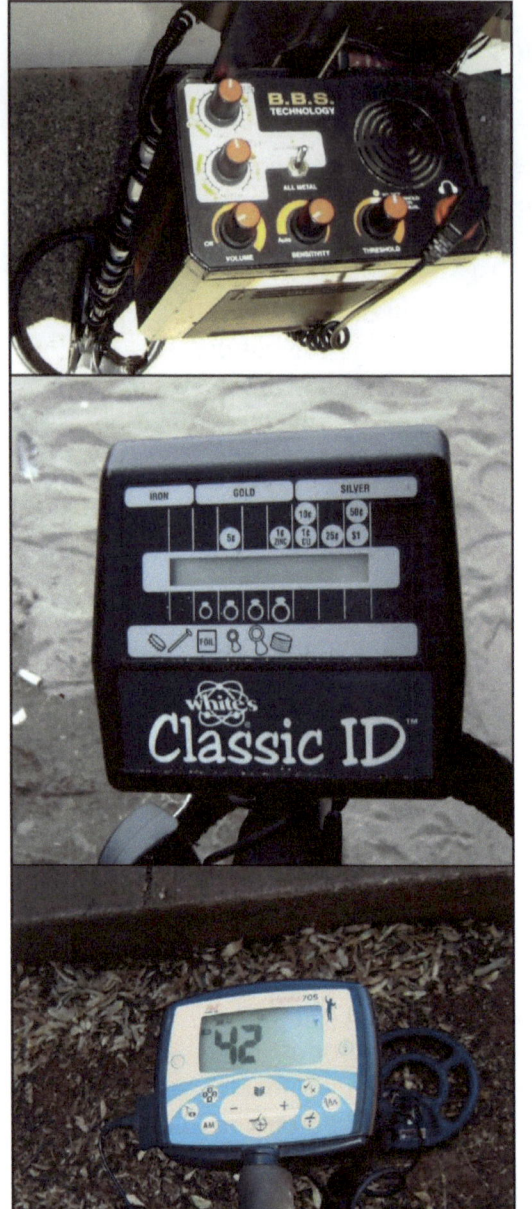

Review:

Figure 22.
Here's a different sound-only detector. Remember, detectors generally link the Phase Shift to the sound you hear, so a junk target makes a low grunt, and a highly conductive silver coin makes a high-pitched tone. Not having a display reduces the costs and sometimes the weight of the detector. This is an older but popular model, the Minelab Sovereign.

Figure 23.
This is another example of a VLF detector with Target ID grouped into ranges, or categories. The categories are shown along the top of the display: from the White's Classic ID detector.

Figure 24.
This is the MineLab X-Terra 705, an advanced metal detector which uses the more accurate VDI numbers to identify the target. Detectors with VDI numbers require more finely tuned circuitry, often resulting in higher costs.

TYPES OF DETECTING

Let's look now at the three types of metal detecting: 1. On **Land**; 2. Beach and **Underwater**; and 3. **Prospecting**.

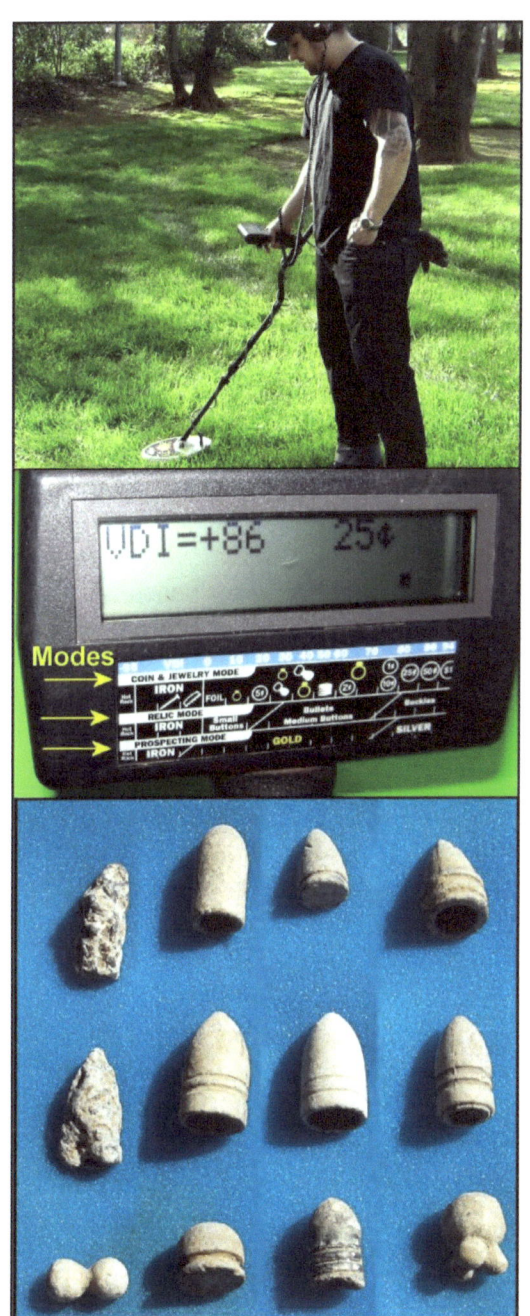

Figure 25.
The most common type of detecting is on land, where you're looking for coins, jewelry and historic artifacts. You may start out just looking for coins but you soon find it's a lot of fun to recover tokens, belt buckles, jewelry, and various types of metal objects.

Figure 26.
There are three Modes (all on land) on the Whites MXT. In practice, many operators adjust the discrimination settings to produce a signal for all or almost all metal objects (All Metal Mode), then use the Target ID indicator to decide whether to dig it up or not.

Figure 27.
Collecting relics – something valued for its age or historic interest – requires a low discrimination level, as many older items were made of iron. Collecting Civil War relics, such as the bullets shown here, is a popular specialty within the land detecting mode.
Photo courtesy of Joe and Phil Gouff.

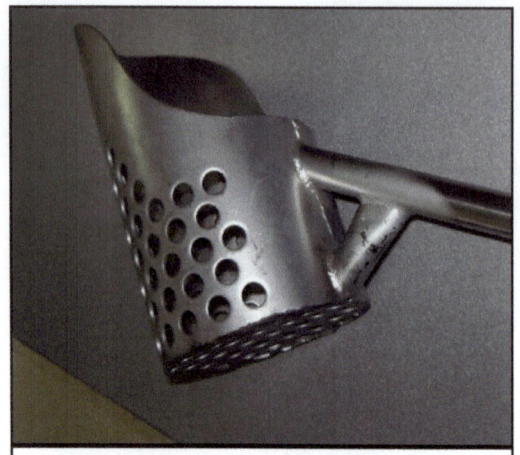

Figure 28.
Beach detecting is another favorite specialty in the hobby. Most detector coils and the attached wiring are waterproof, so you can safely search in shallow water, as long as you keep the control box dry. You will benefit from specialized equipment, such as the sand scoop shown here.

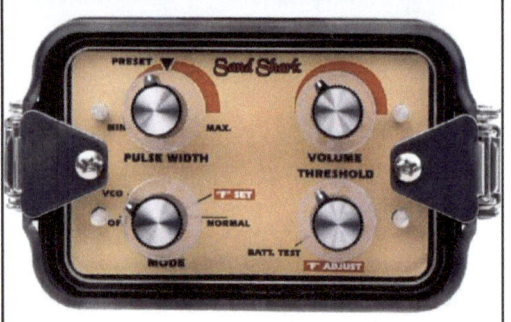

Figure 29.
Several manufacturers produce waterproof metal detectors specifically for beach and water detecting. You can go wading in the ocean, or scan the beaches, or completely submerge waterproof detectors. The Tesoro Sand Shark features waterproof housing.

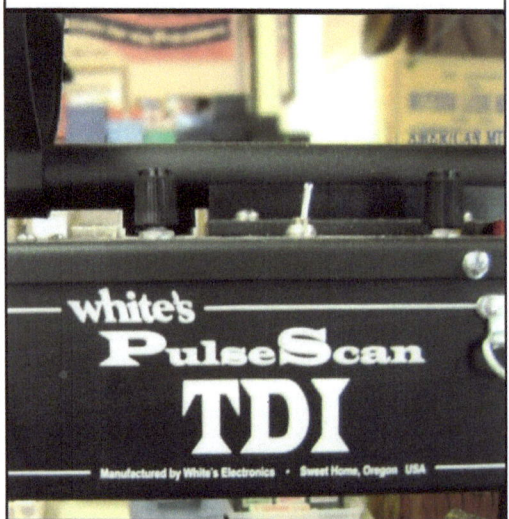

Figure 30.
Some specialized beach detectors use a different kind of circuitry called **Pulse-Induction**, or PI. These PI detectors are relatively expensive and do not use the Phase Shift technology of VLF detectors, so they are not as reliable in Target Identification. They are good at seeing through salty beaches and mineralized soil, as in gold prospecting. The White's Pulse Scan TDI is a PI detector made for gold prospecting.

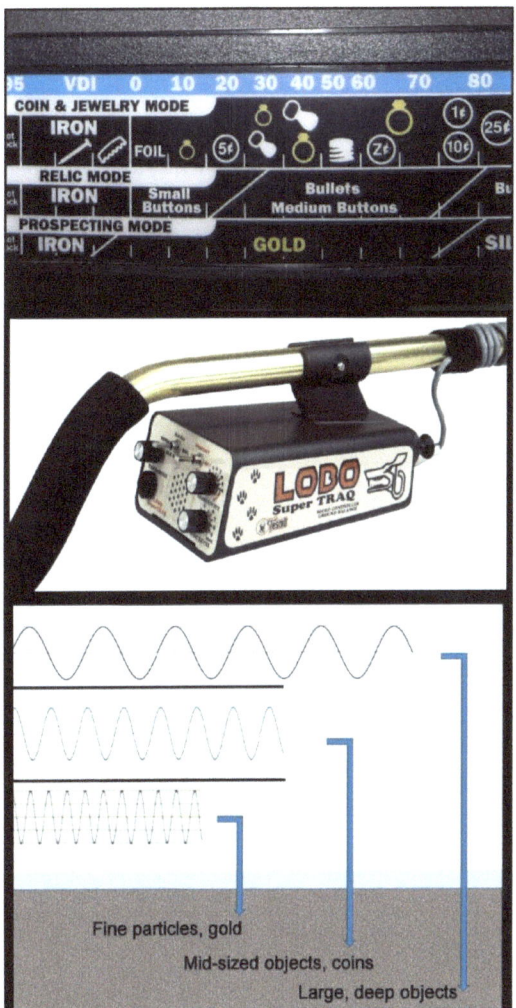

Prospecting refers to searching for any kind of metal (gold, silver, copper), or the ores and minerals that contain them. The vast majority of prospecting, however, in the metal detecting community is for GOLD!

Figure 31.
Since gold occurs in nature at different levels of purity and different mixtures of concentration, the phase shift for gold objects can appear along a wide spectrum on the phase shift dial. You can see that in White's MXT display, which shows gold ranging from about zero to over 70 on the VDI scale.

Figure 32.
Some machines dedicated to gold prospecting will abandon coin identification features altogether, since you're not looking for coins in mines, quarries, or gold fields. Instead they rely on sounds alone, using a low tone for base metals, and a high-pitched "Zip-zip!" sound for gold. The Tesoro Lobo Super Traq is well suited for gold prospecting.

Figure 33.
Most gold occurs in fine particles, so the dedicated gold detector will often employ a higher operating frequency in the VLF range. The higher frequency is better suited to detecting small grains, although it doesn't penetrate as deeply.

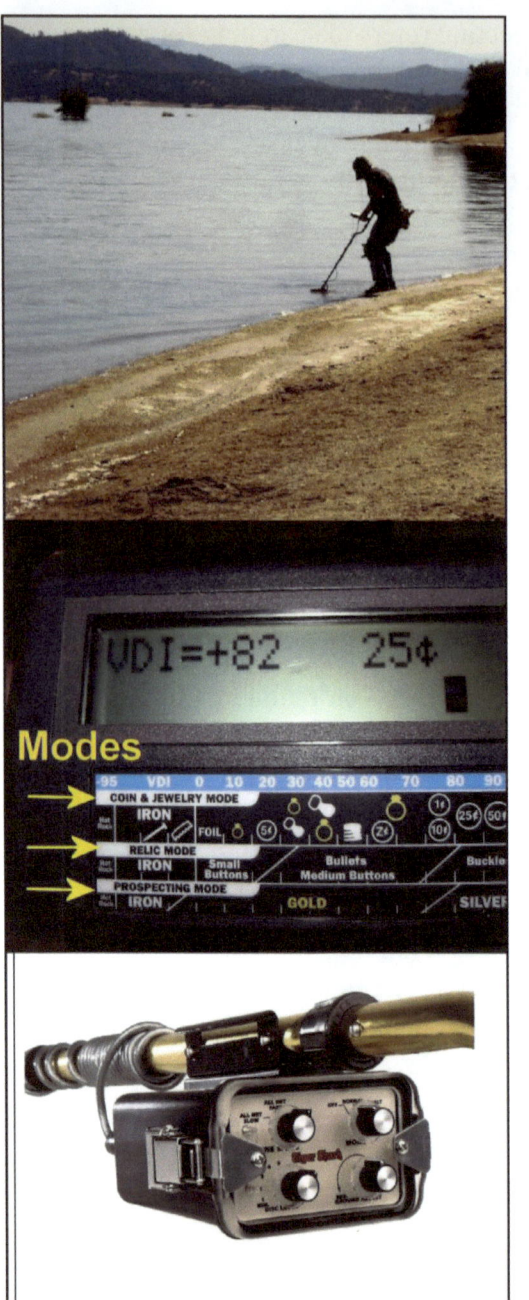

Figure 34.
Some detectors have different Modes, so you can work in more than one area. Land detectors, for example, generally have submersible coils, so you can detect in shallow water.

Figure 35.
This White's detector has a Mode switch, so you can detect for gold as well as for coins. There is another mode setting for relics.

Figure 36.
TEST: What kind of detector is this? Is it for general metal detecting or dedicated to one particular type? How can you tell? Answer: Next page, bottom.

DETECTOR COILS

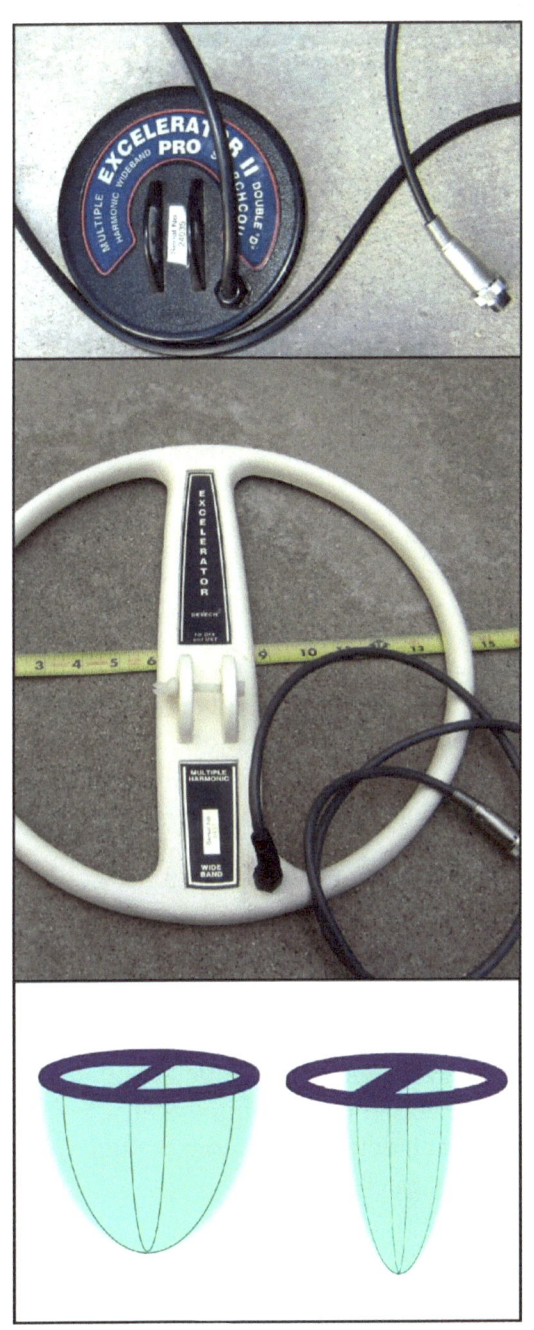

Figure 37.
Detector companies often make different sized coils that are interchangeable with the original coil that comes with the detector. Independent manufacturers sometimes make coils that are compatible with popular brands: One example is the EXcelerator brand of coils.

Figure 38.
Larger coils can detect objects that are deeper than mid-sized coils. Smaller coils are better at telling the difference between two buried objects that are close to each other. This is a 14-inch coil made for White's detectors.

Figure 39.
The "D-D coil" changes the shape of the magnetic search field, from roughly a football shape to that of a flat disk, like a pancake. This allows you to cover more ground with every sweep of the detector, and it helps in pin-pointing the target.

Answer: Figure 36 is a water detector. The name 'Tiger Shark' refers to a common fish. You can see the waterproof controls.

SEARCH TYPES BY REGION

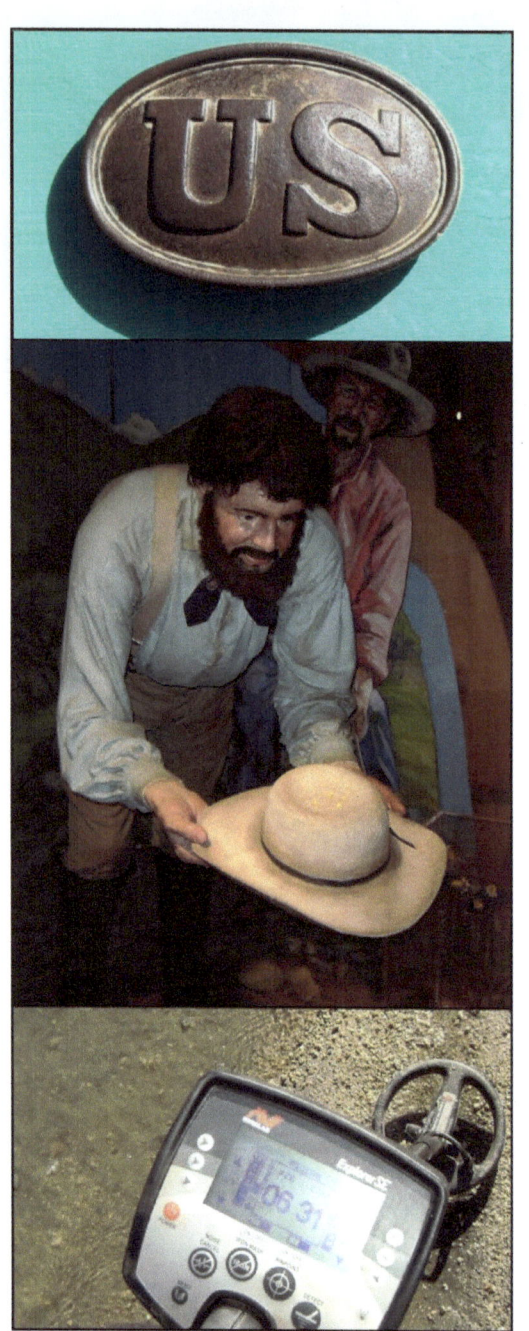

Figure 40.
Relics can be found in all regions of the U.S., but the oldest relics are on the East Coast, where the original 13 colonies were founded. Civil War relics, likewise, are most abundant along the Atlantic Coast states and inland where battles occurred.
Photo courtesy of Joe and Phil Gouff.

Figure 41.
Gold is found in only certain locations in the U.S, such as in California, North Carolina, and Alaska. 23 states have gold deposits. This image is from a display at the Marshall Gold Discovery Park, Coloma, California.

Figure 42.
Lakes and rivers enable water detecting in just about every corner of the United States, although beach detecting is more popular in coastal states, such as Florida.

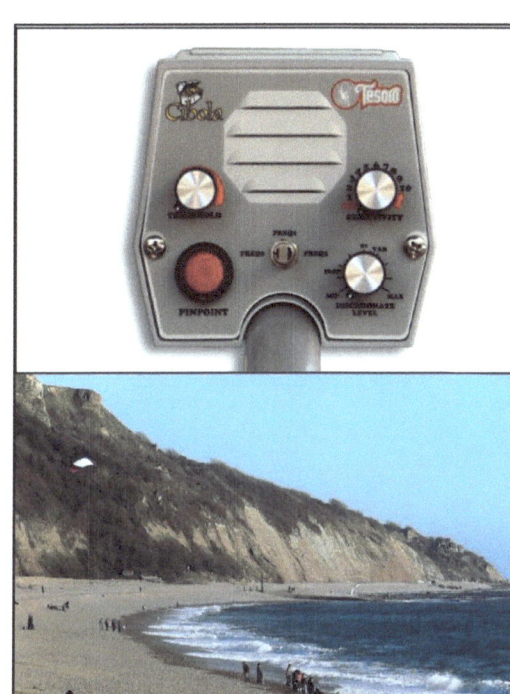

CHOOSING A DETECTOR

Be sure to check out Best Bets for the Beginner, page 25. Then read the reviews from http://metaldetectorreviews.net/.

Figure 43.
Choosing the right detector for your needs depends on a balance between your budget and the features that you feel are important. The street price for the Tesoro Cibola is about $100 more than the Silver Umax. You have to decide if the extra features are worth the extra cost.

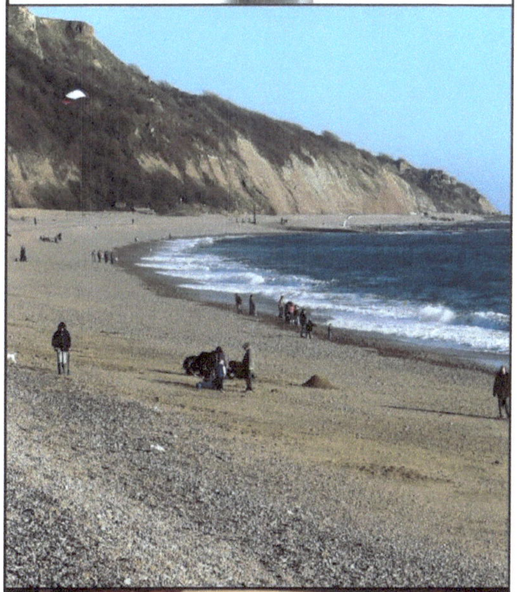

Figure 44.
Choosing a detector should take into consideration the types of detecting available in your area. If you live near a beach, for example, sooner or later you will be out there detecting in the sand.
Photo courtesy of Flickr.com Gazzat.

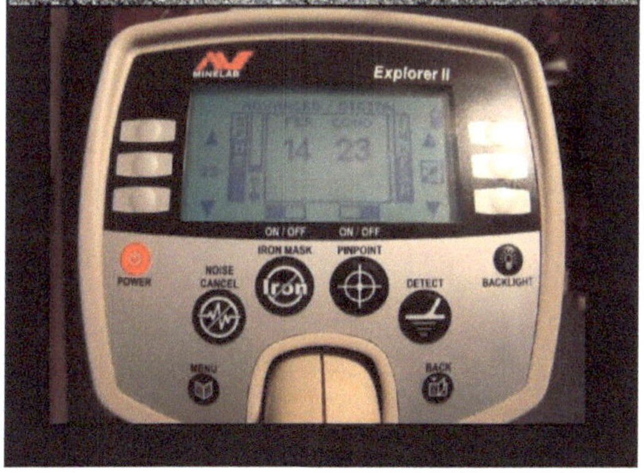

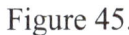

Figure 45.
Buying a used detector may get you more performance at a lower price, but you have to weigh that against the risks of wear and tear on the coil and the control box. The recently discontinued Minelab Explorer II is a hot item in the used detector market due to its advanced features.

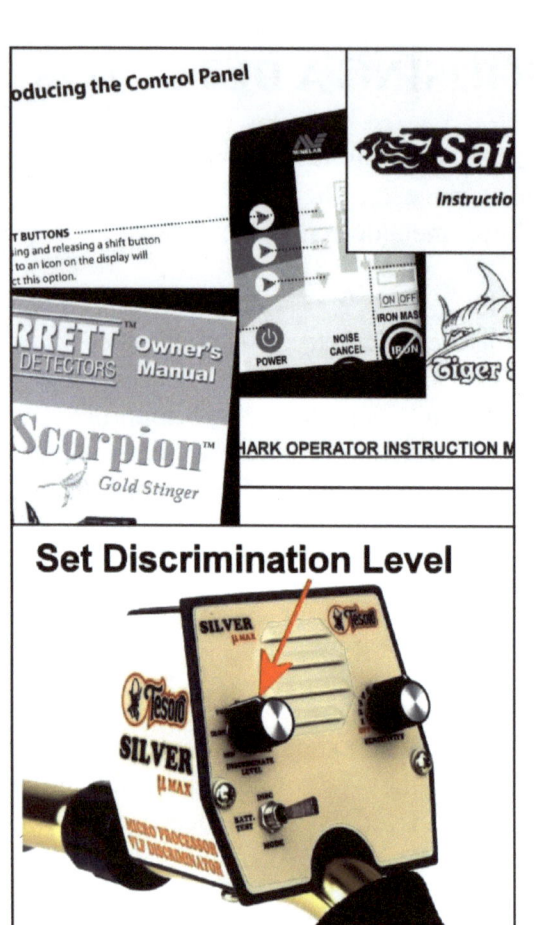

USING YOUR DETECTOR

Figure 46.

The very best source of information on your detector comes from reading the manufacturers' user manual. Be sure to read the instructions thoroughly. User manuals are generally available on line, so you can check out a detector before you buy it.

Figure 47.

The most important adjustment is for the **Discrimination** level, as explained on previous pages. The first thing you'll notice is that you hear lots of junk signals. Scan some test coins so you can tell the difference between trash a real coin.

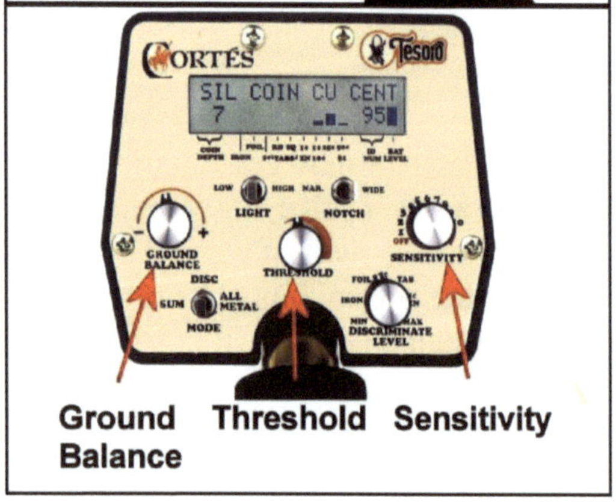

Figure 48.

Most soils have small amounts of metals and salts that can interfere with the signals from coins. **Ground Balance** lets you adjust the rejection of soil minerals. Low cost detectors often have fixed, factory set ground balance. Many detectors have a **Threshold** adjustment and a **Sensitivity** dial. The Threshold should be set to a level where you hear a steady hum in the headphones. The Sensitivity should be set to the highest level that does not produce static or false signals.

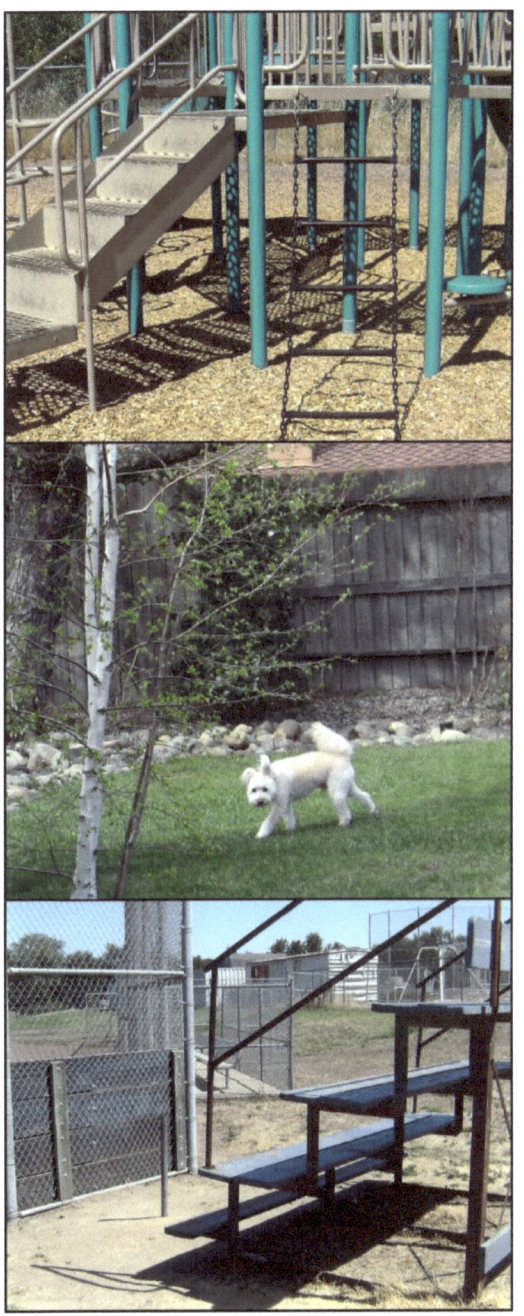

Figure 49.
If this is your first hunt, start out at a "Tot-Lot," which is a playground for children filled with wood chips or other mulch-type material. They are easy to dig and usually contain lots of small metal objects. You'll probably find your first arcade token there.

Figure 50.
For practice, search around your own house. Surveying all the nooks and crannies in your home territory is a good exercise for when you get to search an old homestead.

Figure 51.
Once you've gotten the hang of detecting, check out your local parks. You will easily find a wide variety of coins, jewelry, relics, and trinkets. Here are seats by a baseball field. Do you think people might have dropped things in the dirt beneath the seats?

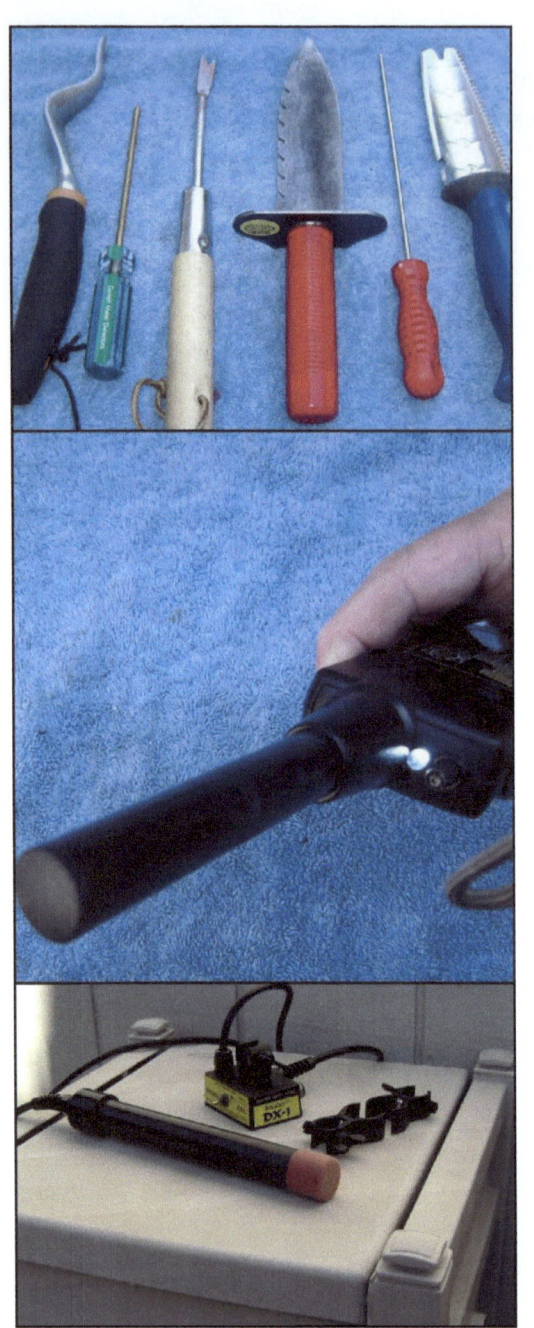

METAL DETECTING ADD-ONS

Figure 52.
You will need some kind of digger to retrieve your finds. Your local soil conditions will guide you to the right tools for your area. This may be something as simple as a screwdriver or a weeding tool, or small hand tools, all the way up to full-sized shovels.

Figure 53.
Hand-held pin-pointers make finding coins a lot faster. Most detectors have a pin-point mode, but the hand-held units can fit easily inside the hole as you dig, making for a speedy retrieval. A White's Bullseye II is shown here. Other popular models are the Garrett Pro-Pinpointer, and the DetectorPro Pistol Probe.

Figure 54.
An **In-Line Probe** is a device that connects to your control box, so the probe retains the Target ID capabilities of your detector, yet it fits right into the area you are digging. This makes it easier to locate coins. At left is a SunRay in-line probe.

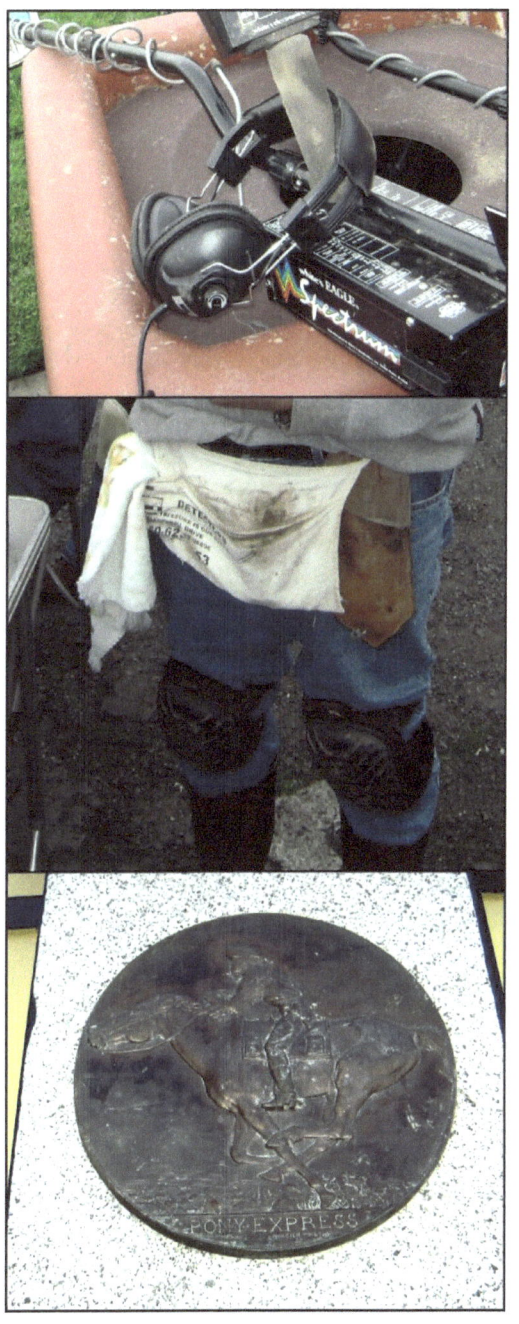

Figure 55.
Headphones save on your battery power. They allow you to hear faint signals and they are a courtesy to the people around you.

Figure 56.
There are all sorts of add-ons that make detecting more comfortable, such as knee pads and a tool belt. Scooping the dirt onto a towel makes it easier to refill the hole, so the lawn looks clean when you leave.

Figure 57.
Historical markers generally mean the site is off limits to detecting. Keep your detecting legal. Be sure to get permission before searching on private property, and know the regulations for digging in parks in your area. Local metal detecting clubs often list park regulations. State and federal historical sites are generally off limits to metal detecting. Check out the Bureau of Land Management (http://www.blm.gov/wo/st/en.html) then click on your state.

HUNTING SKILLS

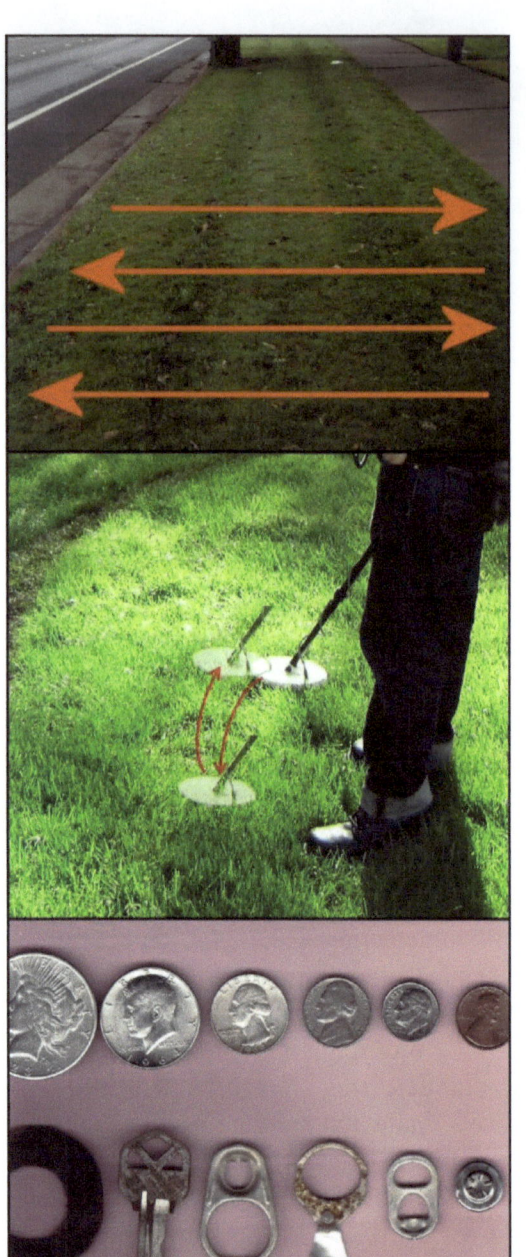

Figure 58.
Get into the habit of searching in a grid pattern or concentric circles, so you're not searching the same area twice.

Figure 59.
Each swing of the detector should overlap the previous one, so the sweep of the coil doesn't miss large areas.

Figure 60.
Pay attention to the sounds your detector makes to common coins. You'll soon learn to distinguish between a coin and junk. This will help you work with your detector's discrimination function, saving you a lot of time digging.

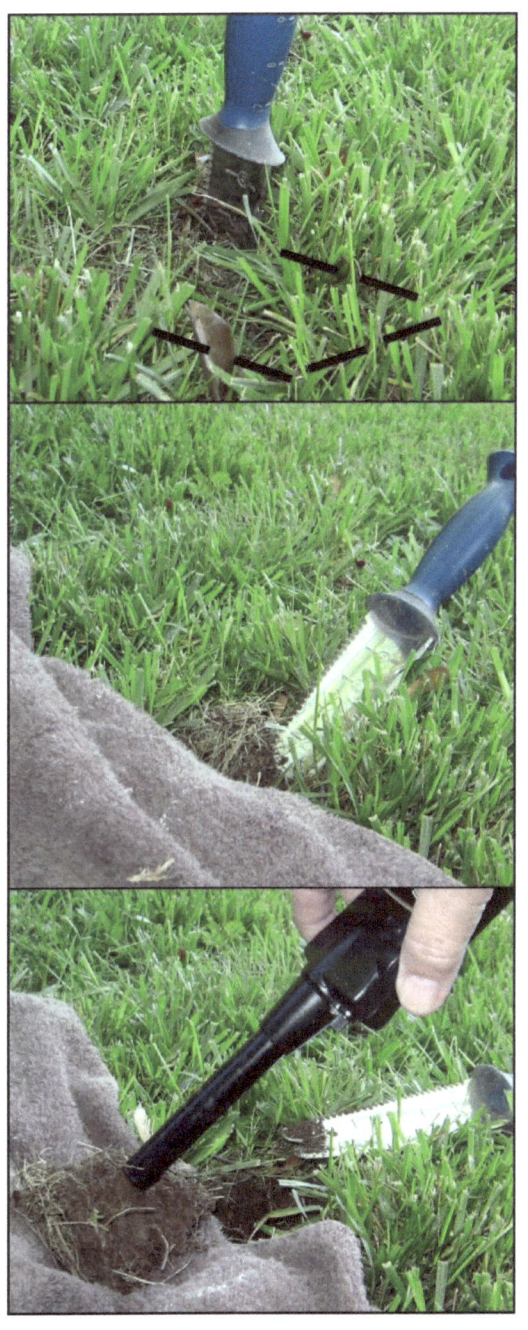

Figure 61.
Keep lawns tidy.

Use the flap technique for retrieving coins. Cut three sides of a square around the target coin using your spade or digging tool.

Lay down a towel and flip the plug onto the towel.

Next search the plug and the hole. Flip the plug back in place. Keeping one side of the plug attached keeps the grass alive. Try to keep the lawn looking as good as it did before you dug the hole.

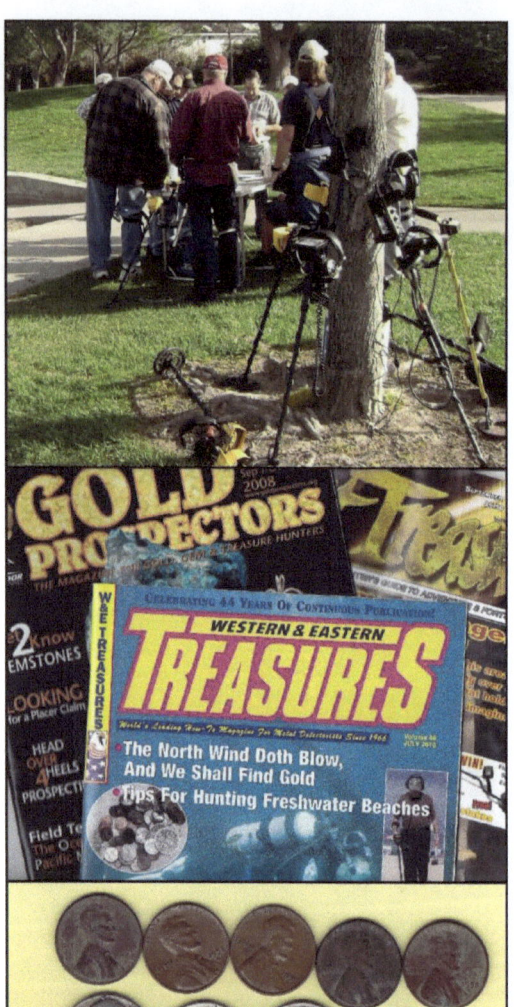

Figure 62.
Consider metal detecting with a partner. It's much safer, mutually beneficial, and you can learn from each other. Metal detecting clubs are a great resource. They often have group hunts, a book and magazine library, and all kinds of experts and friends who can help you

Figure 63.
National clubs and magazines will greatly expand your horizons. If you are considering detecting for gold, take a look at the Gold Prospectors Association of America. See the Resources section at the back of this book.

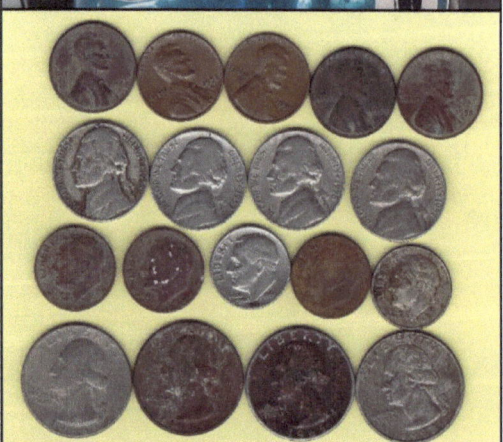

Figure 64.
Don't use chemicals or abrasives to clean your coins! Collector coins can lose substantial value if you treat them harshly. Don't try to remove stains and tarnish. Use mild soapy water to remove surface dirt, and then examine them for their date and collector value. If they have no coin-collector value, you can clean them with a vinegar soak, and then rinse with water.

USING THE INTERNET

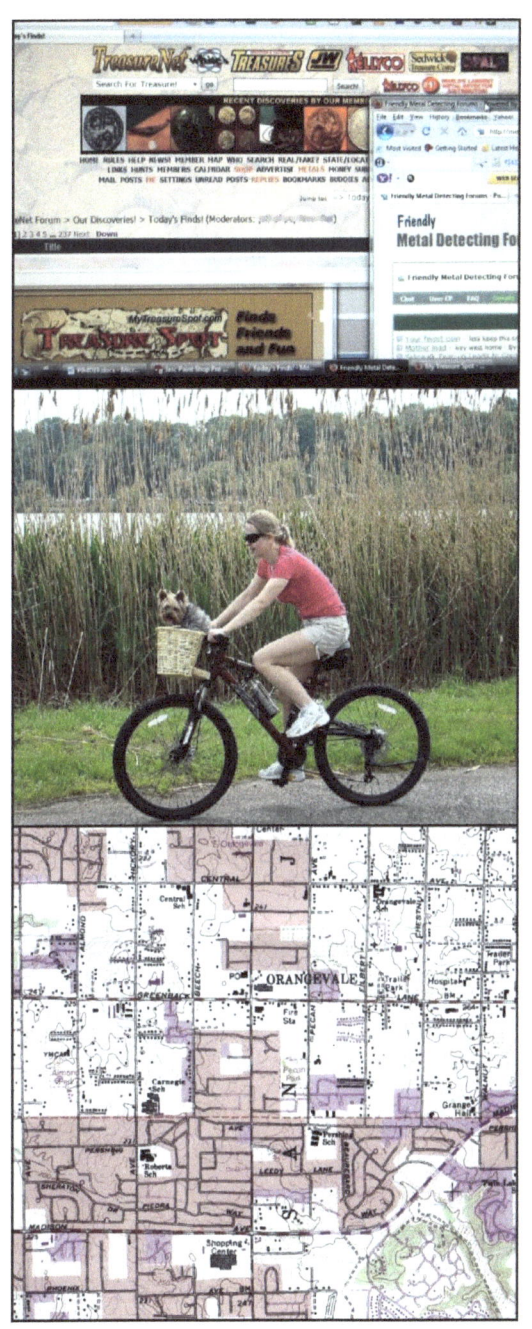

Figure 65.
There is a vast online community of hobbyists who have extensive knowledge of metal detecting. You can learn a lot by participating in forums and discussion groups on the Internet. Again, see the Resources at the end of the book.

Figure 66.
Special interest groups, such as biking enthusiasts, are a great way to research places to hunt. You can look up campgrounds, hiking trails, picnic areas, beaches, and tourist attractions that make good places to hunt.
Photo courtesy of Bobby Mikul, PublicDomainPictures.net.

Figure 67.
The Internet provides access to U.S. Geological Survey Topographic maps, commonly referred to as Topo Maps. They provide lots of detail at a reasonable price. You can look them up online or order a hard copy.

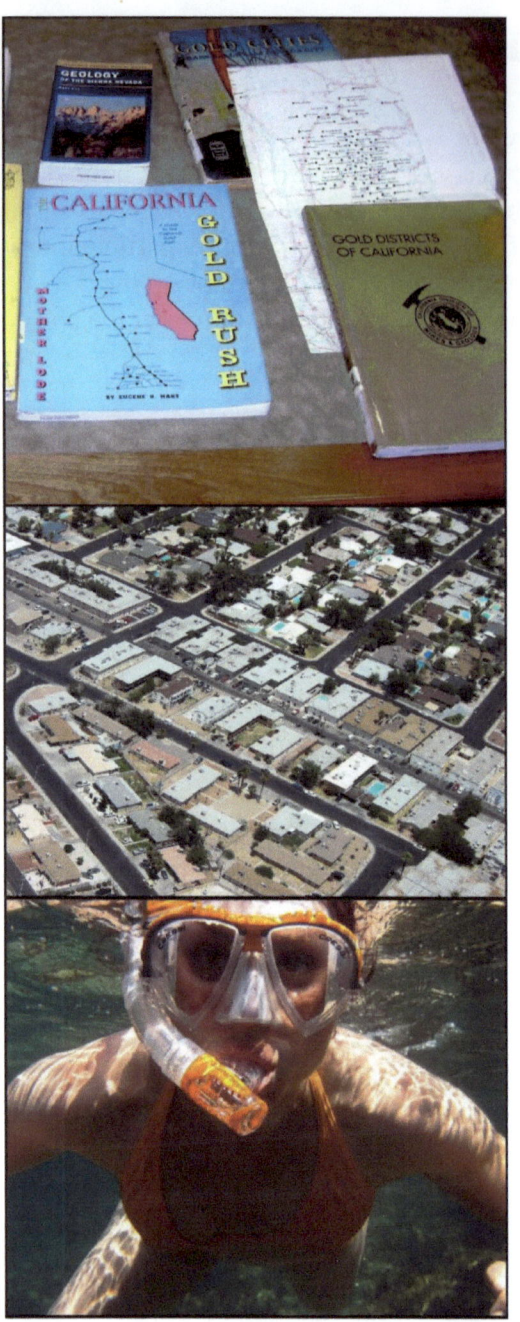

Figure 68.
One of the best ways to find good metal detecting sites is to investigate your local history. Most libraries have a local history section, or you can conduct historic research online. California is fortunate to have a colorful history in gold prospecting.

Figure 69.
Sites like Google Earth and TerraServer.com offer birds-eye and aerial views of search sites in your area. This makes it much easier to select areas to hunt.
Image courtesy of Shari Weinsheimer, from PublicDomainPictures.net.

Figure 70.
The Internet is also a great way to investigate specialties within the metal detecting hobby, such as searching for meteorites, or under-water hunting.
Image courtesy of Petr Kratochvil, from PublicDomainPictures.net.

For more detail on metal detecting, see, *Metal Detecting for the Beginner*, 2nd Edition, by Vince Migliore, available on line.

RESOURCES

A. Best Bets for the Beginner – Detector Choices

There is no "one size fits all" in choosing your first detector. Your best bet is to read the reviews from http://metaldetectorreviews.net/ and other review sites.
- Fisher F2: ~$200; Fast response and sensitive
- Garret Ace 250: ~$210; Popular with good Target ID
- Minelab Musketeer: ~$300; Good depth and discrimination
- Tesoro Silver Umax: ~$250; Fast response, light weight, easy to use
- White's Prism II: ~$250; Good Discrimination, light weight

B. The big players

The following are some of the big players in treasure hunting, as they combine a magazine and a major World Wide Web presence.

TreasureNet
    Internet: http://www.treasurenet.com/
    Forum: http://forum.treasurenet.com/index.php
    Magazine: ***Western & Eastern Treasures***

Lost Treasure
    Internet: http://www.losttreasure.com/
    Magazine: ***Lost Treasure***

Gold Prospector's Association
    Internet: http://www.goldprospectors.org/
    Magazine: ***Gold Prospectors***
    Group: Gold Prospectors Association of America, P.O. Box 891509, Tumecula, CA 92589.

C. Magazines

Western & Eastern Treasures
People's Publishing
P.O. Box 219
San Anselmo, CA 94979
(800) 999-9718

Lost Treasure
LostTreasure
P.O. Box 469091
Escondido, CA 92046
(866) 469-6224

Gold Prospectors
Gold Prospectors Association of America
P.O. Box 891509
Tumecula, CA 92589
(800) 551-9707

American Digger
American Digger
P.O. Box 126
Aeworth, GA 3101
(770) 362-8671

ICMJ's Prospecting and Mining Journal
ICMJ
PO Box 2260
Aptos, CA 95001
(831) 479-1500

D. Online resources

These are the top metal detecting Internet sites, in order by popularity using the alexa.com web traffic statistics:

- TreasureNet: http://www.treasurenet.com
- Kellyco: http://www.kellycodetectors.com/indexmain.php
- The Friendly Metal Detecting Forum: http://metaldetectingforum.com/index.php
- Lost Treasure On Line: http://www.losttreasure.com
- Find's Treasure Forums: http://www.findmall.com/
- Metal Detector.com: http://www.metaldetector.cc/index.asp
- Treasure Quest: http://www.treasurequestxlt.com/
- Treasure Depot: http://www.thetreasuredepot.com/index.html
- Metal Detector Reviews: http://metaldetectorreviews.net/
- Metal Detecting World: http://metaldetectingworld.com/
- OKM, Germany: http://www.okmmetaldetectors.com/
- Go Metal Detecting: http://gometaldetecting.com/
- Treasure Hunting: http://www.treasurehunting.com/

E. Major brands

The major brand metal detector manufacturers generally sell detectors for land based coin, jewelry and treasure hunting, as well as specialty detectors for prospecting or underwater searching. They often sell supplies and peripherals for hobbyists.

In alphabetical order:

Bounty Hunter
1465 Henry Brennan Dr # H
El Paso, TX 79936
(915) 633-8354
(800) 444-5994
Web: http://www.detecting.com/
Bounty Hunter sells its products through major retailers, such as The Bounty Hunter Store (http://www.BountyHunterStore.com), and Kellyco, (http://www.kellycodetectors.com).

Garrett Electronics
1881 W. State Street
Garland, TX 75042
Tel: (972) 494-6151
Fax: (972) 494-1881
Web: http://www.garrett.com/
Email: sales@garrett.com
Garret has a large security division in addition to the hobby division. Charles Garrett has written several books on metal detecting.

Fisher Labs
1465-H Henry Brennan
El Paso, TX 79936
Tel: (915) 225-0333
Fax: (915) 225-0336
Web: http://www.fisherlab.com/hobby/index.html
Email: info@fisherlab.com
Fisher Labs has three divisions, hobby, industrial, and security.

JW Fishers Mfg Inc
1953 County Street
East Taunton, MA 02718
Tel: (800) 822-4744
Web: http://www.jwfishers.com/
Email: Info@jwfishers.com

Minelab USA
871 Grier Dr., Suite B1
Las Vegas, NV 89119 USA
Tel: (702) 891-8809
Fax: (702) 891-8810
Web: http://www.minelab.com/usa/consumer
Email: info@minelabusa.com

Teknetics
1465-H Henry Brennan
El Paso, TX 79936
Tel: 1-800-413-4131
Web: http://www.tekneticst2.com/

Tesoro Electronics
715 White Spar Road
Prescott, AZ 86303
Tel: (928) 771-2646
Web: http://www.tesoro.com/
Email: support @ tesoro.com

White's Electronics
1011 Pleasant Valley Road
Sweet Home, OR 97386
Tel: (800) 547-6911
Fax: (541) 367-6629
White's also has regional offices around the US: See
http://whiteselectronics.com/info/contacts.html for local offices and suppliers.

F. Suppliers

Suppliers, in alphabetical order:

Aardvark Metal Detectors (Distributor)
1085 Belle Avenue
Winter Springs, FL 32708
Web: http://www.aardvarkdetectors.com
Email: sales@aardvarkdetectors.com
Tel: (800) 828-1455

Accurate Locators (Manufacturer)
1383 2nd Ave.
Gold Hill, Oregon 97525
Tel: (877) 808-6200
Web: Web: http://www.accuratelocators.com/

DetectorPro (Distributor)
Web: http://www.detectorpro.com/
Email: info@detectorpro.com
Distributor for Headhunter metal detectors and "innovative treasure hunting concepts."
See their CyberStore at http://www.detectorpro.com/cyberstore/cyberstore1.htm

Doc's Detecting Supply (Distributor)
3740 S. Royal Crest Street
Las Vegas, Nevada 89119
Web: http://www.docsdetecting.com/
Email: cop704@yahoo.com
Tel: (800) 477-3211 Ext. 14
Distributor for Coiltek brand coils for Minelab detectors.

Famous Treasures (Distributor)
Tampa Florida
4312 Land o' Lakes
Land O' Lakes, FL 34639
Toll Free (888) 788-1819
(813) 996-1787
Email: sales@famoustreasures.com
Website: http://www.famoustreasures.com/

Jimmy Sierra Products (Accessories)
James and Jim Normandi
6880 Sir Francis Drake Blvd. (P.O. Box 519)
Forest Knolls, California 94933
Telephone: 1-800-457-0875
Website: http://www.jimmysierra.com/
jimmsierra@jimmysierra.com

JW Fishers Manufacturing (Manufacturer)
1953 County Streets
East Taunton, MA 02718
Web: http://www.jwfishers.com/
Email: info@jwfishers.com
Underwater detectors - Note: Different from Fisher Labs.

Kellyco (Distributor)
1085 Belle Ave
Winter Springs, FL 32708
Tel: (888) 535-5926
Web: http://www.kellycodetectors.com/indexmain.php
Email: orderdept@kellycodetectors.com

Note: Kellyco and other "superstore" distributors carry lesser-known brands and specialty items such as:

Automax (Pinpointing probe)
Link: http://www.kellycodetectors.com/vibra/automaxprecisionpinpointer.htm
Cobra (Metal detector manufacturer)
Link: http://www.kellycodetectors.com/cobra/cobramain.htm
MP Digital (Metal detector manufacturer)
Link: http://www.kellycodetectors.com/MP3/MP3information.htm
Nautilus (Metal detector manufacturer)
Link: http://www.kellycodetectors.com/nautilus/nautilus.htm
Nokta Engineering
Link: http://www.kellycodetectors.com/nokta/nokta_buy.htm
Pioneer (Bounty Hunter)
Link: http://www.kellycodetectors.com/bountyhunter/pioneer_main.htm
Teknetics (Metal detector manufacturer)
Link: http://www.kellycodetectors.com/Teknetics/teknetics.htm
Titan (Metal detectors)
Link: http://www.kellycodetectors.com/titan/titan.htm
Viper (Metal detector manufacturer)
Link: http://www.kellycodetectors.com/cobra/vipersmain2.htm

Outdoor Outfitters (Distributor)
705 Elm Street,
Waukesha WI 53186
Web: http://www.outdoorout.com/
Email: Outdoorout@ameritech.net
Tel: (800) 558 2020
Fax: 262 542 4435

Predator Tools (Digging tools)
35 South Woodruff Road
Bridgetown, NJ 08302
Web: http://www.predatortools.com/
Web: sales@predatortools.com
Tel: 856-455-3790
Fax: 856-455-6604

Simmons Scientific Products (Locating rods)
P.O. Box 10057
Wilmington, NC 28404
Web: http://www.simmonsscientificproducts.com/
Email: simmonssp@aol.com
Tel. & Fax: (910) 686-1656

Sunray Detector Electronics (In-line target probes, distributor)
106 N Main Street
P.O. Box 300
Hazleton, Iowa 50641-0300
Web: http://www.sunraydetector.com/
Email: infor@sunraydetector.com
Tel: (319) 636-2244

Notes:

www.ingramcontent.com/pod-product-compliance
Lightning Source LLC
Chambersburg PA
CBHW040753200526
45159CB00025B/2084